A Bee or Not a Bee?

Activity Book

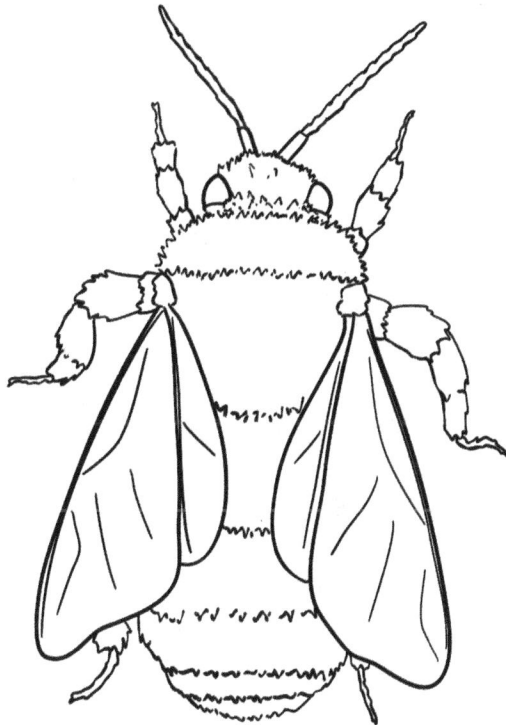

Written & illustrated by Audrey Sauble

Colored by _____

Do you ever wonder about bees? What is a bee?

Honeybees are one type of bee.

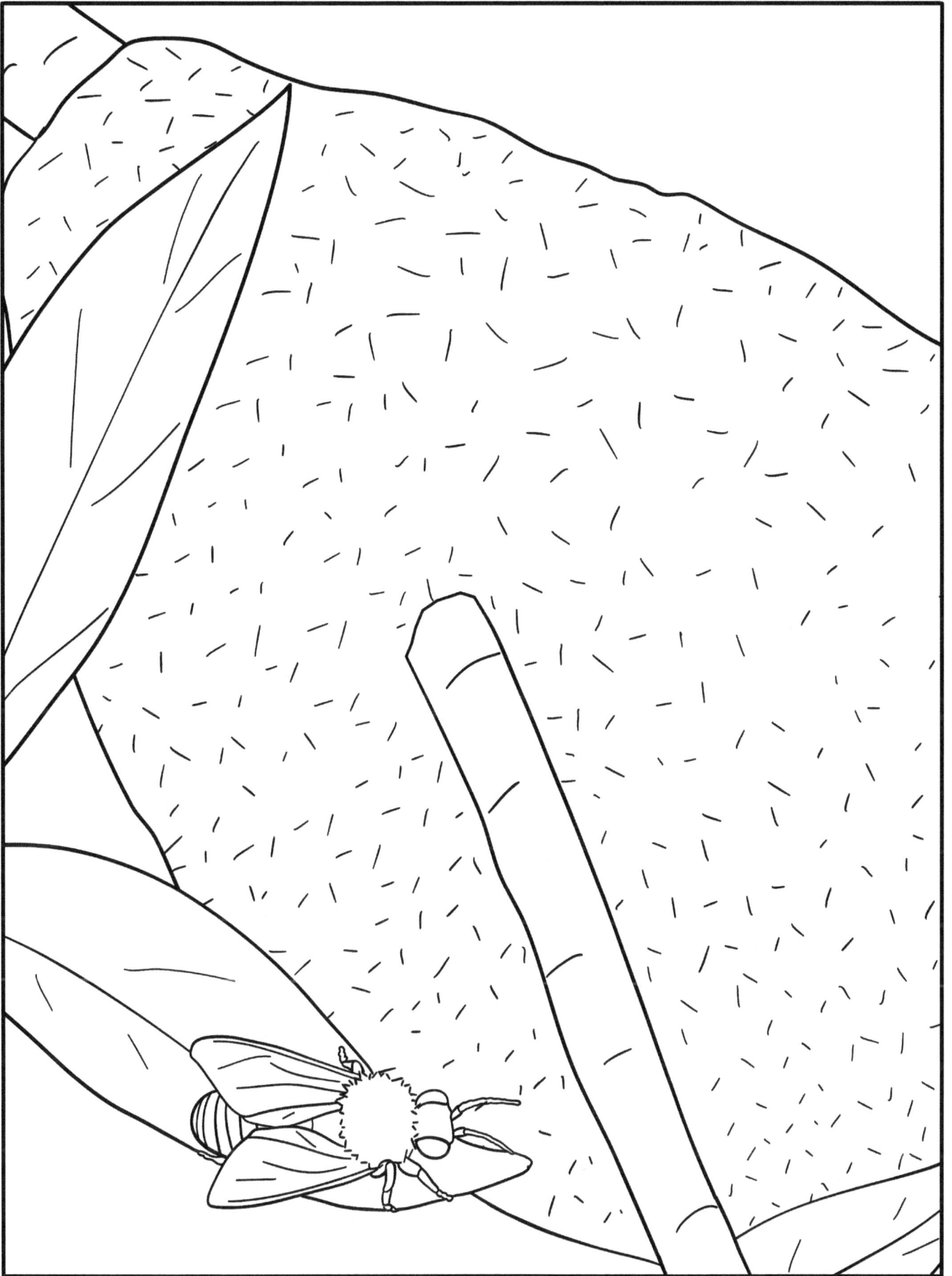

Honeybees are fuzzy insects with stripes. Honeybees make honey and pollinate flowers.

Bumblebees also gather pollen and nectar, but they don't make honey.

Some flies also pollinate plants. But these flies aren't bees, and they don't make honey.

A bee is an insect with six legs. It does not have eight legs like a spider.

A bee egg hatches into a larva or grub, not a miniature adult the way some insects do.

A bee has clear wings, not leathery wings like a beetle.

A bee has four wings, not two like a robber fly.

A bee has a narrow face like an ant, not a round face like a drone fly.

A bee has long, bendy antennae, not short, poky ones like a hover fly.

A bee has wings as an adult, not just when it swarms as ants do.

Most bees have thick, hairy legs, not narrow legs like wasps, ants, and sawflies.

Bees feed their young nectar and pollen rather than meat like wasps do.

Some bees are social bees. These bees nest together in hives.

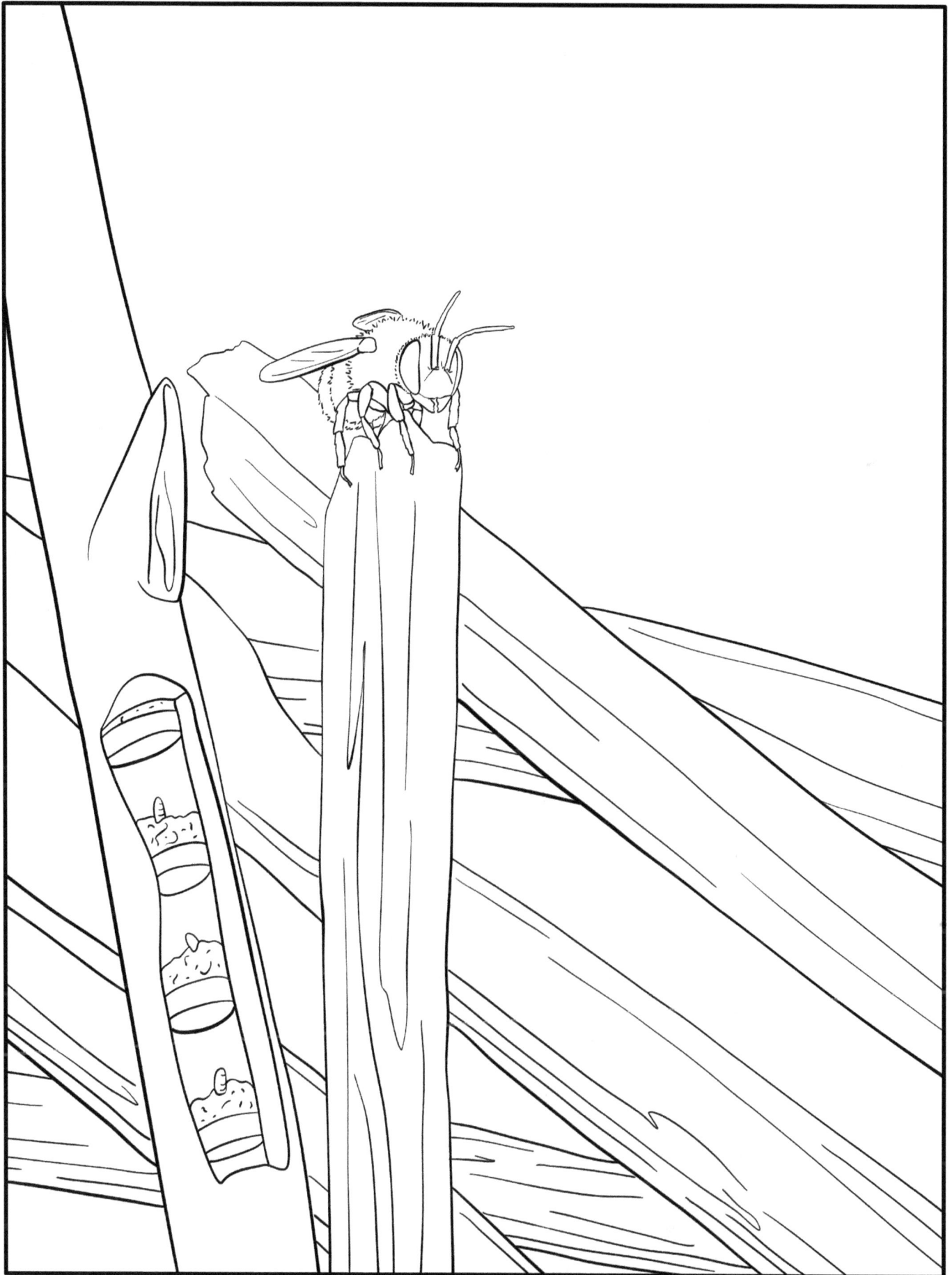

Most bees are solitary bees and nest alone. Some solitary bees nest in hollow stems.

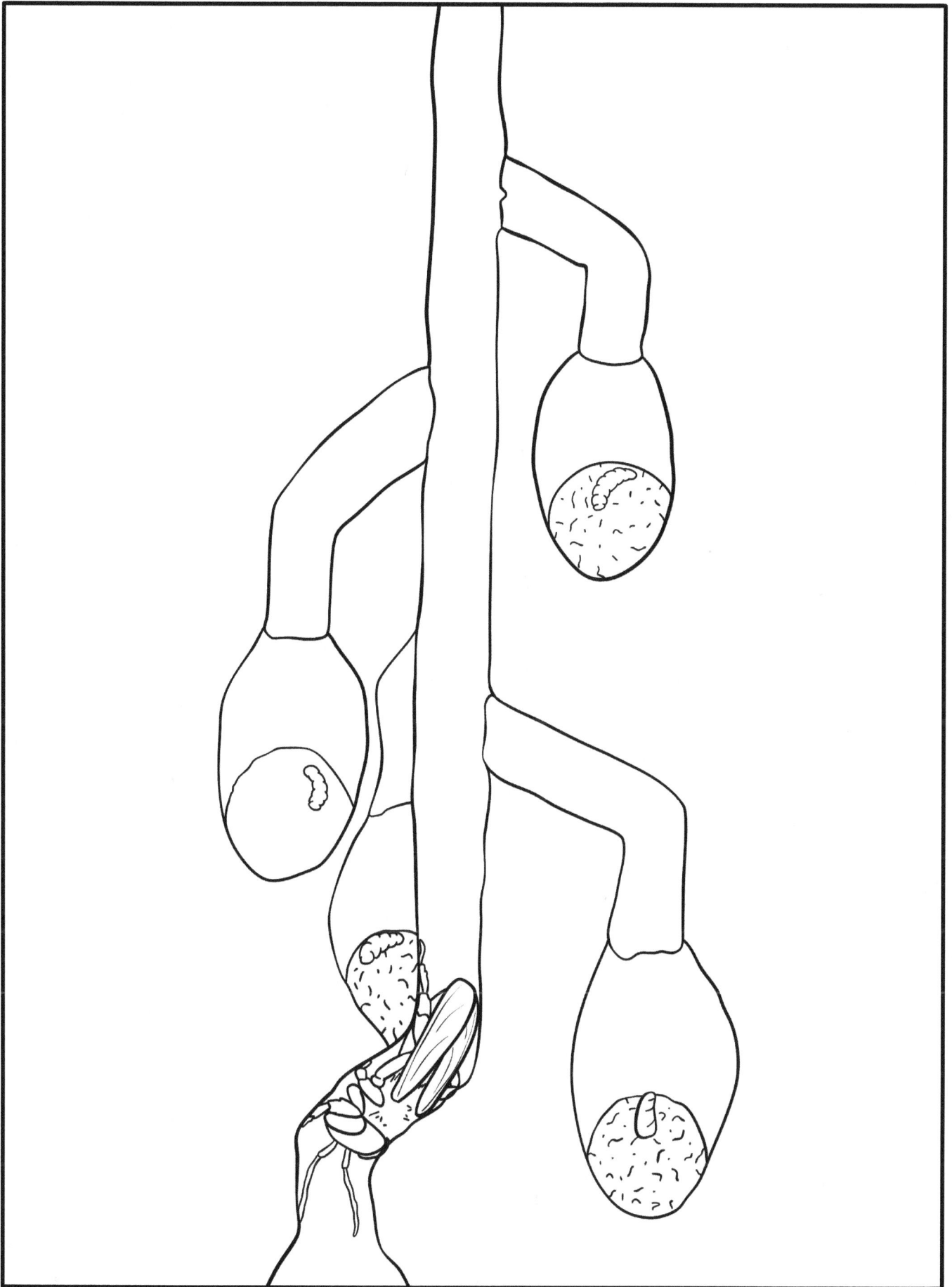

Other solitary bees dig tunnels for their nests.

A bee can be many colors and sizes, but all bees are amazing.

Just like honeybees, these bees are all bees.

Can you guess what kind of flower this is? Cut out these rectangles and arrange them like puzzle pieces to find out. Then glue the pieces onto another piece of paper. If you want, you can also write a sentence explaining how bees and flowers help each other.

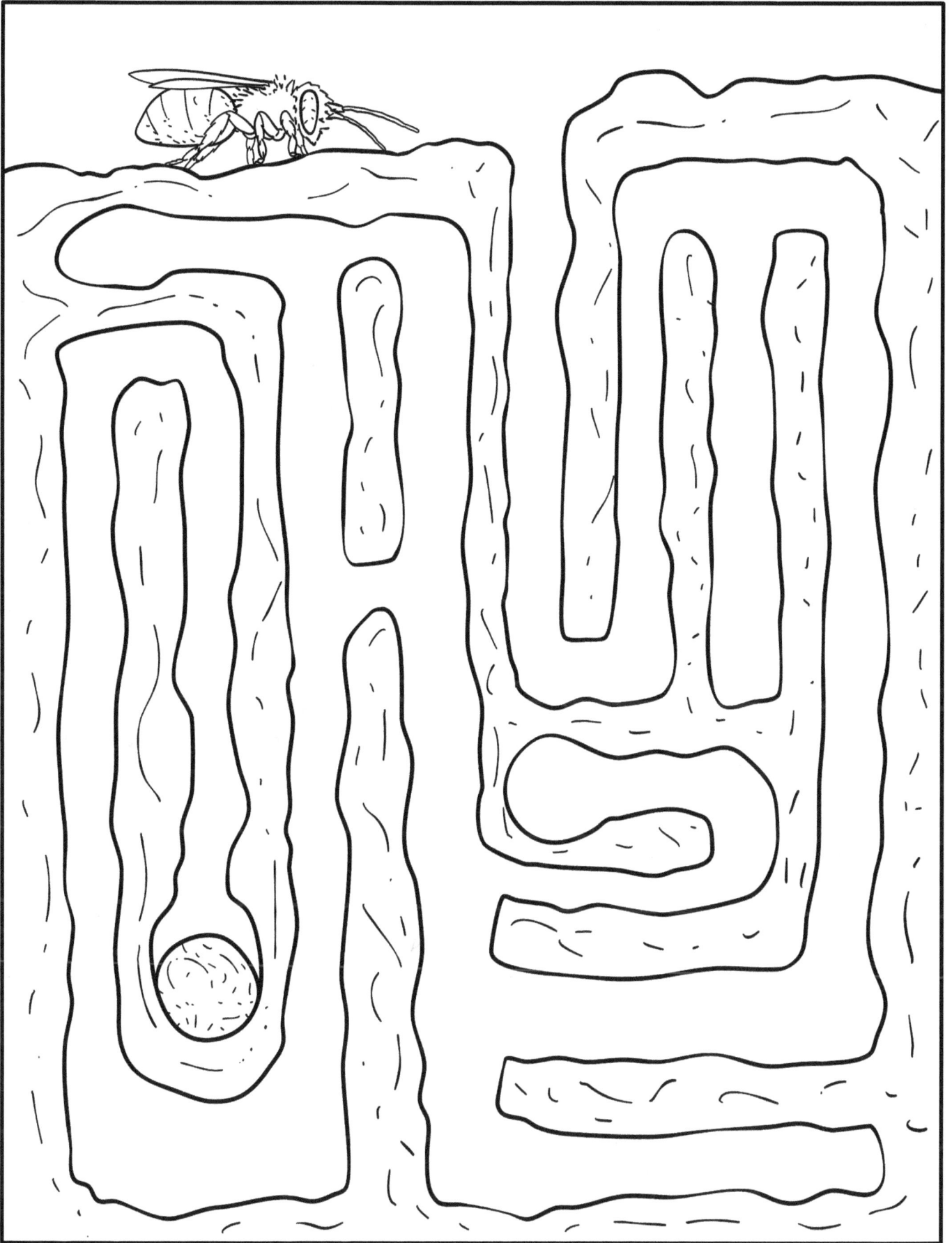

Help the miner bee find her way back to the ball of pollen waiting in her nest so she can lay her egg on the pollen.

Can You Unscramble
These Words?

1. NOEEYEBH

2. EULEBMBEB

3. NILLATEOP

4. WOERFL

5. NAPLTS

6. ILORATYS

7. AICOSL

8. ENTS

9. NNAENAET

10. OOLCNY

11. IHEV

12. YLINGF

P	C	L	I	N	A	A	G	E	T
C	V	V	L	E	G	S	L	H	M
L	H	A	K	L	K	I	B	O	N
O	S	O	B	O	D	Y	T	N	F
N	U	E	E	S	L	P	S	E	E
H	I	V	E	T	O	U	S	Y	L
A	Q	F	F	I	N	S	E	C	T
I	G	D	M	N	E	S	T	S	U
R	M	R	B	G	K	T	J	O	W
P	M	X	S	W	I	N	G	S	Q

Word list: BEE, HONEY, HIVE, NEST, INSECT, STING, HAIR, LEGS, WINGS, BODY

Big or small, yellow or brown—we have lots of words to describe bees, but the most important ones for identifying bees include legs, hair, and wings. Look across and down to find the hidden words.

U	S	B	U	D	P	Y	O	N	A
S	X	P	N	L	E	A	F	G	R
T	B	W	E	T	U	J	L	N	J
E	R	S	E	E	D	S	O	E	U
M	A	R	T	P	Q	V	W	C	A
O	N	O	L	E	M	E	E	T	I
M	C	O	A	T	N	R	R	A	L
T	H	T	F	A	R	I	T	R	A
G	O	S	E	L	E	A	J	R	M
G	P	O	L	L	E	N	X	J	P

Word list: PETAL, LEAF, STEM, NECTAR, BRANCH, POLLEN, BUD, FLOWER, ROOTS, SEEDS

Plants have many important parts. Can you find the names for these plant parts in this word search? Look across and down to find the hidden words.

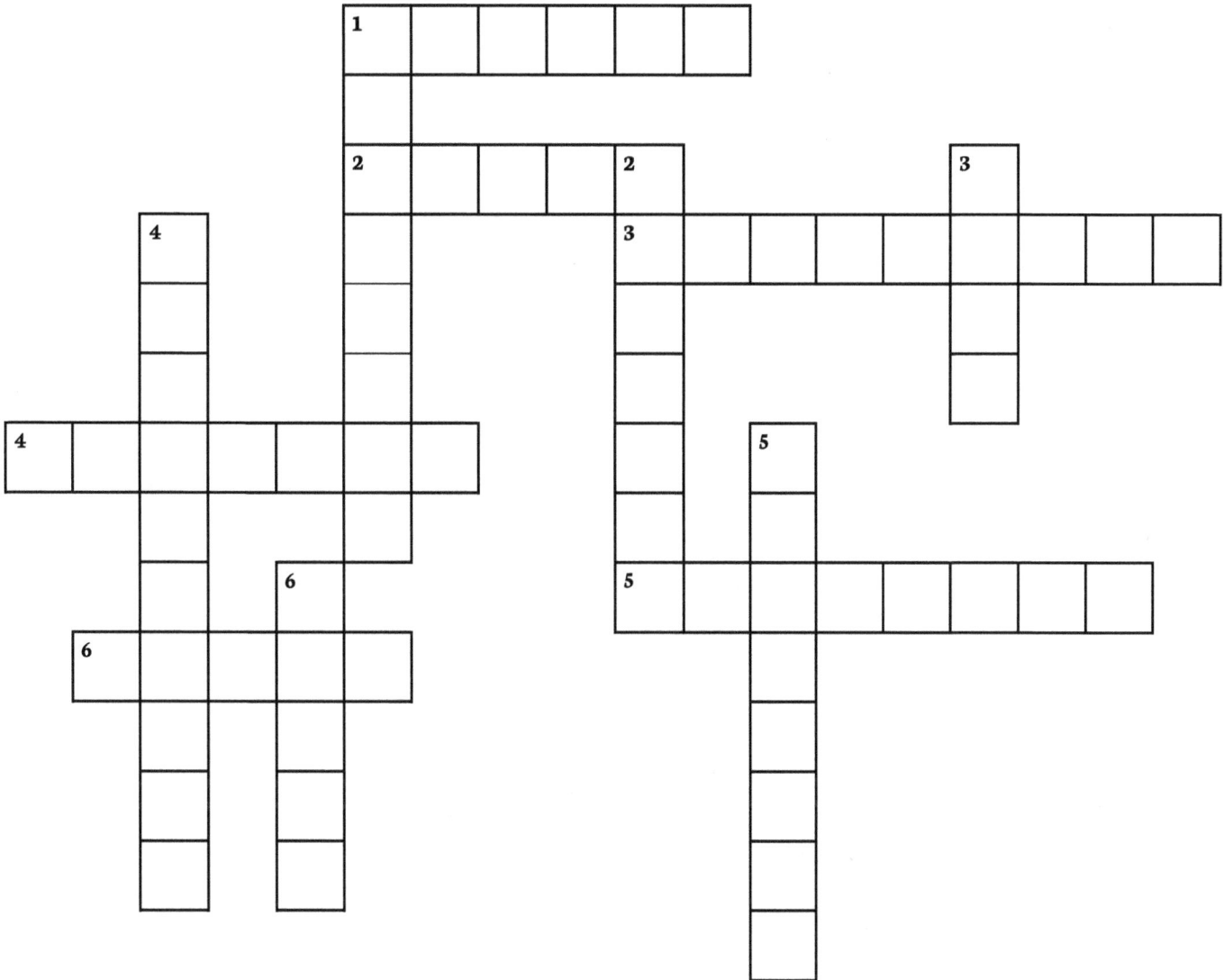

Across:

1. Some bees work together. These bees are _____ bees.

2. Some _____ look like bees, but they feed their young meat not pollen.

3. Bees help _____ flowers.

4. _____ have strong elytra or outer wing.

5. Some bees live alone. These bees are called _____ bees.

6. _____ have only two wings.

Down:

1. _____ are closely related to bees and wasps and may look similar.

2. _____ are not insects; they have eight legs.

3. Queen _____ lose their wings when they are ready to build a nest.

4. _____ hover around plants and help pollinate flowers, but they aren't bees.

5. Only honeybees form large _____.

6. Bees make _____ to lay their eggs.

1 = Purple   3 = Light blue   5 = Yellow

2 = Green   4 = Orange   6 = Dark gray

Can you guess what this bumblebee's favorite flower is? Color any spaces without a number black, and then color each numbered section in this picture with the correct color to find out.

Name: _____ **Date:** _____

Design a Science Project

My goal is to: _____

_____

_____

First: _____

_____

Then: _____

_____

_____

I will need: _____

_____

_____

_____

Results: _____

_____

_____

_____

_____

Need some ideas? Try one of these projects: design a bee house, plan a garden with wildflowers, run a bee survey to see how many different kinds of bees you can spot.

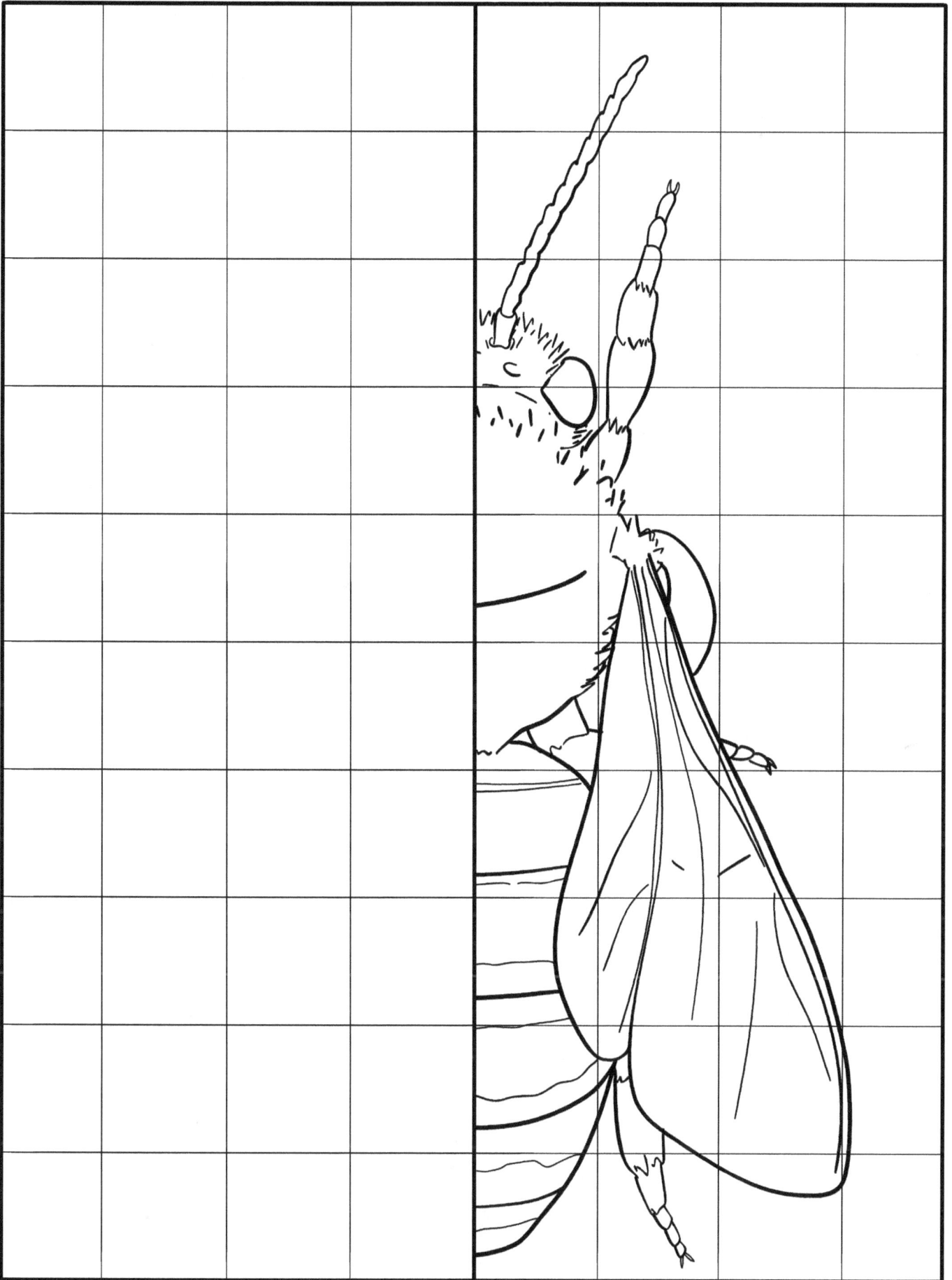

Use the grid as a guide and draw the other side of this picture. Then color the bee.

Mason Bee House

For this project, you will need an adult's help.

Start with a large block of untreated wood.

With a drill, drill a series of holes into the wood. The holes should be ⅜ to ¼-inch wide and 6 inches deep. They don't need to be in rows, but be careful not to make the holes too close to each other.

When your mason bee house is finished, mount it in a sunny area several feet above the ground. Make sure it sheltered so that rain doesn't drip into the nests.

Benefits:

- Adding mason bee houses supports native solitary bees. While honeybees get the most attention, these solitary bees are also important pollinators and need help too.

- Solitary bees are less likely to sting than honeybees, so it may be safer to add mason bee houses than a honeybee hive.

Potential challenges:

- When many bees gather in a mason bee house, it's easier for pests and diseases to spread to the other bees. It may be necessary to replace mason bee houses every year or to use the type of house that can be opened and cleaned.

Other Options:

- If you don't want to make your own bee house, you can buy one instead. Many of these commercially built houses are designed to be easy to clean as well.

- You can also help the bees by doing less yard work. Native bees nest in old plant stems and tiny hollows in old logs or twigs, as well as in open patches of dirt. You could help bees by leaving some dead stems on your plants or saving some of the trimmings in a sheltered corner of your yard.

- You could even ask your school or a local nature center to add a mason bee house! Be ready to explain why solitary bees are an important part of our ecosystem.

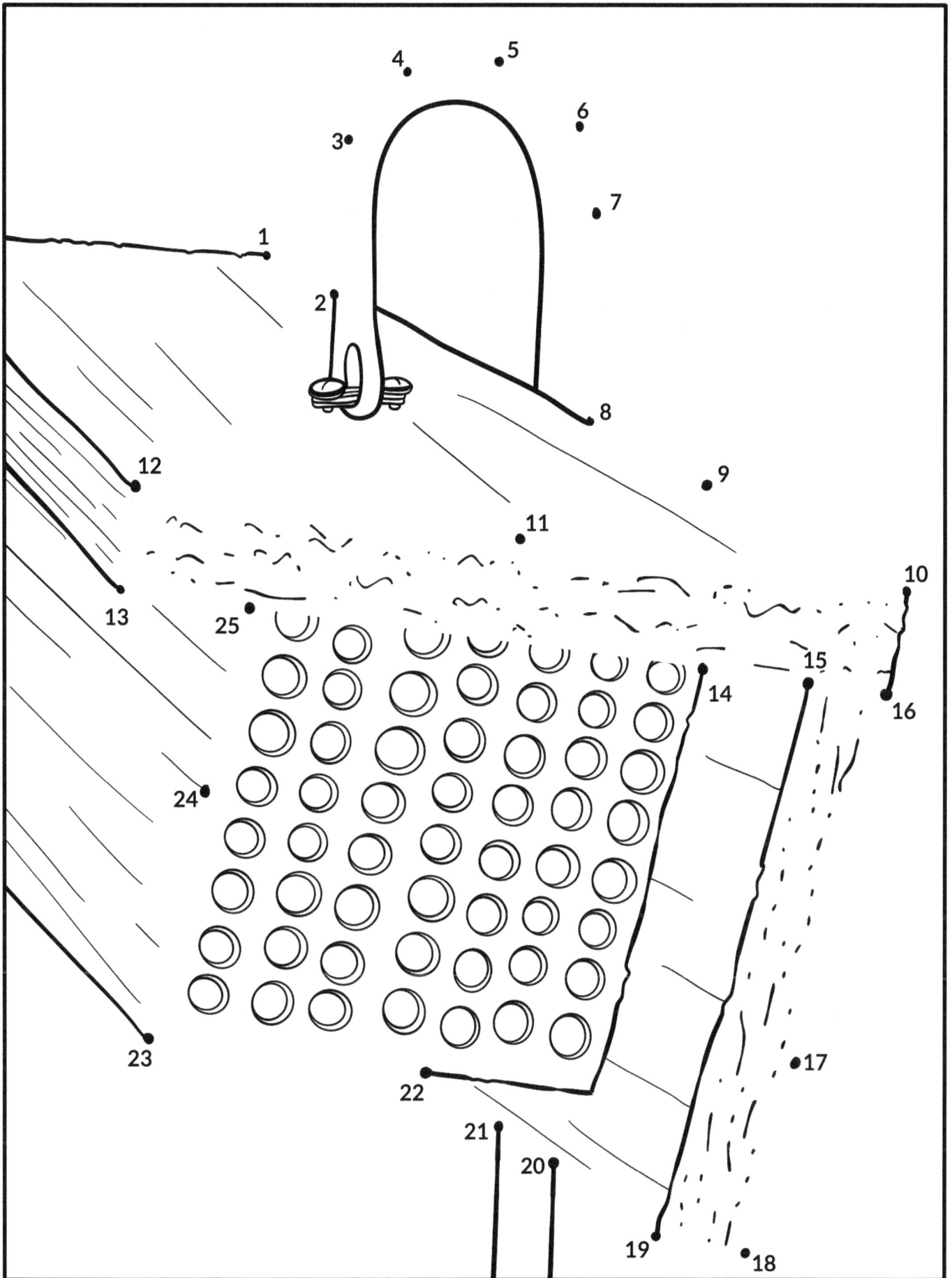

Connect the dots and color the picture to find out what this bee house look like.

Name: _____ **Date:** _____

Insect Hunt

Can you find all of these insects? Use this list to keep track of the ones you've seen. If you need help identifying any of these insects, the iNaturalist website or app may be able to help—but be sure to ask an adult for help first.

Date found:

□· Honeybee _____

□· Bumblebee _____

□· Green or blue bee _____

□· Tiny bee _____

□· Longhorn bee _____

□· Wasp _____

□· Hoverfly _____

□· Green or blue fly _____

□· Fruit fly _____

□· Large ant _____

□· Small ant _____

□· Striped beetle _____

Bee Trivia

1. If a creepy-crawly has six legs, it is an _____.

2. If a creepy-crawly has eight legs, it is not an insect, but it might be a

    _____.

3. All bees have _____ wings.

4. True or false: honeybees are the only insects who pollinate flowers.

    _____.

5. Flies are not bees, even if they have yellow stripes. Flies have

    _____ wings, not _____ wings.

6. While mason bees may nest close together (especially in a mason bee
    house), each bee makes its own nest and works alone to collect the pollen its
    larvae need. Mason bees are _____ bees.

7. True or false: all bees make honey. _____.

8. While they can be very similar to bees in other ways, _____
    shed their wings when they are ready to start a nest.

9. If an insect's wings are clear, it could be a bee, but if it's wings are tough
    and leathery, it's probably a _____.

10. Wasps feed their young _____ (including other insects), but
    bees feed their larvae _____.

Did You Know?

Most people are familiar with western honeybees, but *Melipona beecheii* are a species of stingless honeybees from Central America. These bees belong to the same family as honeybees, but their colonies are much smaller than honeybee colonies. Ancient Mayans called this bee *xunan kab* or regal bee lady (*see page 15*).

Solitary bees aren't as well-known as honeybees, but they are still important pollinators. They're just harder to use for farming because they nest alone, instead of in large colonies like honeybees. Scientists are now studying whether some mason bees can be used for farming. These bees like to nest in mason bee houses, so farmers can encourage mason bees to pollinate their crops by placing mason bee houses near the crops during the spring (*pages 33, 57*).

Even bees that are closely related to each other can vary drastically in size. Furrow bees in the genus *Halictus* are group of ground nesting bees. These bees range from small, shiny bees to large bees the size of honeybees (*page 21*).

All bees use their antennae for gathering information, but some bees have unusually long antennae. Longhorn bees have antennae that can be as long as the bee's body. (*page 17*).

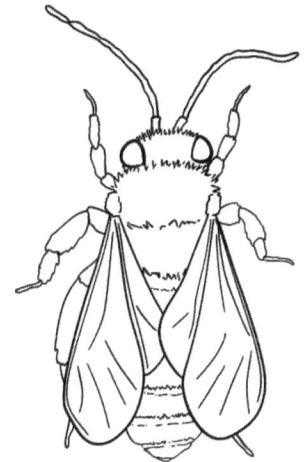

While many bees dig nests in the ground, some bees have unique ways of protecting their nests. Cellophane or plasterer bees are a type of mining bee that produce a clear, waterproof material. The bees use this material use to line their nests to keep their eggs dry (*page 13*).

It's your turn! Do you have a favorite bee? Draw it here and write about it below.

More Activities

Model a bee. Use clay or modeling dough to build a model of a bee. Make sure to include all of the bee's body parts: head, thorax, abdomen, six legs, four wings, and two antennae.

Create a cardboard or paper model of a bee house. You could create a realistic bee house like the mason bee house, or you could design something new from your own imagination.

Plant a flower garden. You could turn this into a research project by looking up native wildflowers from your region and finding out which ones bees like. You make a simple container garden with some marigolds or a lavender bush in a container outside. Or you could even create an indoor garden with paper flowers using tissue paper and pipe cleaners.

Find a bee survey to join. The Great Sunflower Project (greatsunflower.org) is one option, but you may be able to find a local other bee research project on the iNaturalist or Scistarter websites. These surveys ask citizen scientists (like you) to photograph bees and upload the photographs to the survey's website. Then scientists use these photos to identify different bee species. *For this project, you will need an adult to help you look up information on the Internet and decide whether it's okay to share your photos or not.*

Recommended Picture Books:

Honeybee: The Busy Life of Apis Mellifera by Candace Fleming
The Bumblebee Queen by April Pulley Sayre
Give Bees a Chance by Bethany Barton
If Bees Disappeared by Lily Williams
The Thing About Bees: A Love Letter by Shabazz Larkin

To find more book recommendations and activity pages, visit www.aesauble.com/bees.

Answer Key

Word Search #1 grid:

```
P  C  L  I  N  A  A  G  E  T
C  V  V  L  E  G  S  L  H  M
L  H  A  K  L  K  I  B  O  N
O  S  O  B  O  D  Y  T  N  F
N  U  E  E  S  L  P  S  E  Z
H  I  V  E  T  O  U  S  Y  L
A  Q  F  F  I  N  S  E  C  T
I  G  D  M  N  E  S  T  S  U
R  M  R  B  G  K  T  J  O  W
P  M  X  S  W  I  N  G  S  Q
```

Page 49: Word Search #1

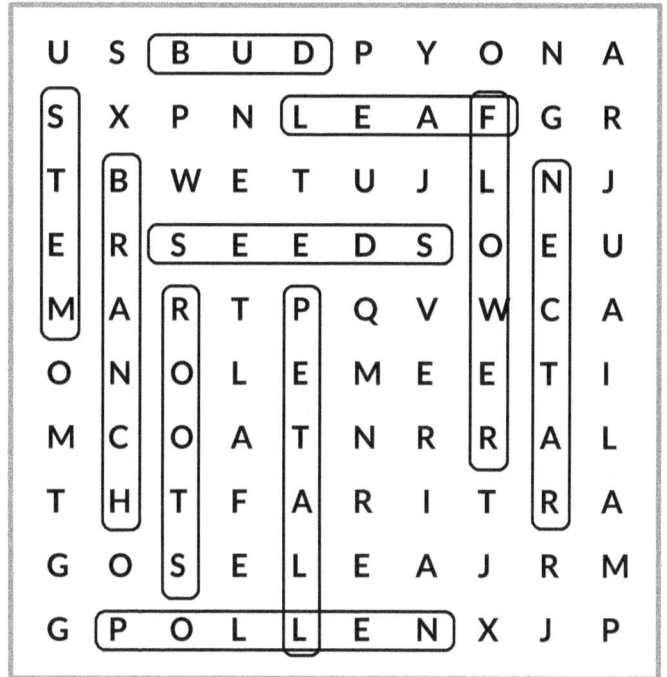

Word Search #2 grid:

```
U  S  B  U  D  P  Y  O  N  A
S  X  P  N  L  E  A  F  G  R
T  B  W  E  T  U  J  L  N  J
E  R  S  E  E  D  S  O  E  U
M  A  R  T  P  Q  V  W  C  A
O  N  O  L  E  M  E  E  T  L
M  C  O  A  T  N  R  R  A  A
T  H  T  F  A  R  I  T  R  A
G  O  S  E  L  E  A  J  R  M
G  P  O  L  L  E  N  X  J  P
```

Page 49: Word Search #2

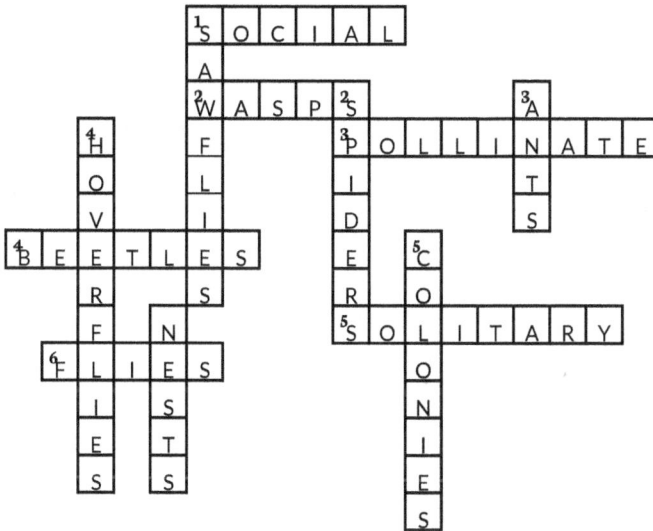

Crossword Puzzle grid:

```
1S O C I A L
 A
 2W A S P   2S        3A
4H   F       3P O L L I N A T E
 O   L       I        N
 V   I       D        T
4B E E T L E S   5C   S
 E   R       E   O
 R   N       R   L
 F       N   5S O L I T A R Y
6F L I E S       O
 I       S       N
 E       T       I
 S       S       E
                 S
```

Page 51: Crossword Puzzle

Page 47: Unscramble the Words

1. Honeybee; 2. Bumblebee; 3. Pollinate; 4. Flower;
5. Plants; 6. Solitary; 7. Social; 8. Nest; 9. Antennae;
10. Colony; 11. Hive; 12. Flying.

Page 65: Bug Trivia

1. insect; 2. spider; 3. four; 4. false; 5. two, four;
6. solitary bees; 7. false; 8. ants; 9. beetle; 10. meat, pollen.